YOUR KNOWLEDGE HAS VALUE

- We will publish your bachelor's and
  master's thesis, essays and papers

- Your own eBook and book -
  sold worldwide in all relevant shops

- Earn money with each sale

Upload your text at www.GRIN.com
and publish for free

Björn Linnemann

Innovation Regions in Europe

GRIN Verlag

Bibliografische Information der Deutschen Nationalbibliothek:

Die Deutsche Bibliothek verzeichnet diese Publikation in der Deutschen National-
bibliografie; detaillierte bibliografische Daten sind im Internet über http://dnb.d-
nb.de/ abrufbar.

Imprint:

Copyright © 2010 GRIN Verlag GmbH
Druck und Bindung: Books on Demand GmbH, Norderstedt Germany
ISBN: 978-3-656-25203-0

This book at GRIN:

http://www.grin.com/en/e-book/198589/innovation-regions-in-europe

Verfasser:

Björn Linnemann

M Sc Geographie

Institut für Geographie – Universität Hamburg

Innovation Regions in Europe

Hauptseminar

New Geographical Spaces:

Globalization and Dynamics of Metropolitan Regions

Hamburg, 11.01.2010

CONTENTS

1 Introduction

Ever since Schumpeter's work on the business cycles, innovation is widely claimed to be a central aspect of economic development. With market expansion reaching its limits and mass production being relocated into low cost countries, the role of innovation gained more and more importance in order to maintain growth in western industrial countries. In contrast to the industrial processes, innovation is not a process, which can be standardised or routinised. It does not follow mathematically ascertainable schemes, but is rather a dynamic process, which is formed by several factors interacting and enforcing each other (Cooke and Morgan, 1998). The aim of this paper is to shed light on empirically determined factors that promote innovation in certain regions. Eventually some remarkable examples of innovative regions in Europe will underline and verify their importance.

2 Spatial aspects of innovation – Why innovations occur in clusters

2.1 Sharing of tacit knowledge

As a result of technological progress, information and knowledge from all over the world is virtually available everywhere, but still, with the sector of long distance business travel growing steadily, there must be reason for seeing each other (Storper and Venables, 2005). In this context it is important to highlight, that different kinds of knowledge do exist and that this is only the case for codified or explicit knowledge. On the other hand, there is knowledge, which is not codified yet or nearly impossible to be codified, the tacit knowledge (Kujath, 2009).

Skills can hardly be transported by description and certain knowledge cannot be articulated due to inadequacies of language. Some skills simply cannot be transported from one individual to another with explaining, but rather need to be shown. Exemplarily the master-apprentice-principle with observation, imitation, correction and repetition seems to be the way to go (Nonaka, 1991). Lundvall et al describe it as a process of doing-using-interacting (DUI) in a community with skilled people (Lundvall et al., 2007). Figure 1 shows the principles of the learning processes. Consequently one can say, that codified knowledge is global and tacit knowledge is local. Learning certain values follows the same principle. They need to be learned and internalised and they are crucial in order to act as member of a community and gain knowledge from it. Sometimes even the skilled worker him- or herself might not know about his or her skills to the full extend. So the tacit knowledge might not even be obvious (Takeuchi and Nonaka, 1995). Thus it is not easily shared and therefore a rare good, which is important for success (Gertler, 2007).

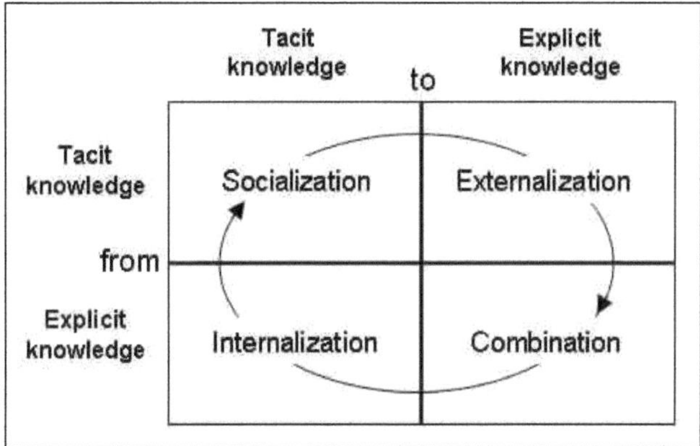

Figure 1: Knowledge loop by Takeuchi/Nonaka
source: Nonaka/Takuchi, 1995

Learning is an interactive, socially embodied and localized process. To gain tacit knowledge or even transform into general knowledge embodied in new and innovative products, it needs embeddedness of the learner in a local community (Lundvall et al., 2007). Being a part of a creative system with a lot of input and creativity is a fertile ground for further development of own ideas, receiving support and knowledge about what the market accepts or desires (Gertler, 2007). Following this logic, tacit knowledge is the prime determinant of the geography of innovative activity. A diversity of tacit knowledge tends to concentrate in metropolitan regions, which serve as gateways of functions and processes and thus attract people from different regions and backgrounds. There one finds universities and other scientific institutions and a stable and rich resource of skilled people from different cultural backgrounds. This cultural and educational diversity leads to a formation of a brain pool and a high density of knowledge available in metropolitan regions and as a consequence the knowledge spillovers occur. Breschi and Malerba see an empirical proof of new knowledge occurring more efficiently among geographically closely related actors (Breschi and Malerba, 2005). Audretsch and Feldman support this statement, when they say, that as a consequence, production of innovation has the tendency to happen in clusters, where knowledge inputs are available (Audretsch and Feldman, 1999). Knowledge simply rubs off on people in innovative places (Storper and Venables, 2005). People might not even be aware of the learning, just by interacting with people of certain skills and capabilities (Maskell, 2005). So it is obvious that knowledge can only effectively be transmitted via interpersonal contacts, which is eased by close geographical and cultural proximity (Breschi and Malerba, 2005). Thus it requires spatial and relational proximity to share tacit knowledge and to create leading edge technology. This is the case for high-tech knowledge as well as low-tech knowledge like skills in management, logistics, sales, marketing, in fact all the day-to-day business drivers, which are crucial for success. Consequently, proximity is the precondition for effective transmission of tacit knowledge (Amin and Cohendet, 1999). A concentration of certain knowledge or skills within a region acts as an attractor for skilled people or specialists from other regions and countries. They move to the area in order to gain from the abundance of knowledge and at the same time contribute to the knowledge pool. So the facilitation of further growth of the cultural and knowledge assets in a region is inherent to regions with such characteristics. This accumulation of knowledge, by generating new knowledge, keeping knowledge and attracting new knowledge to the region makes a region a flourishing place of innovation (Milberg et al., 1996). This phenomenon is evidence, that geographical and relational proximity has advantages, which otherwise would hardly be available (Cooke, 2005). One needs to be an insider in order to know certain firms and consider cooperation and ventures or gain from knowledge. Thus, those lacking this connection do not even see the potential and the opportunity (Sorensen, 2005). Even if knowledge spillovers happen locally and spread into the wider area or even the world in time, far away distances are subject to a time lag, which leads to remote participants being less competitive (Jaffe et al., 1993).

2.2 The role of metropolitan regions

The metropolitan region is an area formed by a cluster of metropolitan functions independent from administrative boundaries. These regions are not formed by administrative boundaries, but rather reach to suburbs and neighbouring cities, villages and areas in regards of the connection of certain factors like infrastructure, services or commuting of labour. They are sources and outlet of material and immaterial streams such as goods, money or knowledge (Grabow and Becker, 2009). While serving as nodal points, metropolitan regions carry strategic functions, such as being center for innovation and competition, a center for decision and control, a gateway with major traffic knots like highways, major train stations, airports, harbours and other factors regarding logistics. They are center of finance, communication with trade fairs or congress centers for example and center of cultural diversity, which form due to a high share of migrants. There are pull factors like jobs, political stability and cultural openness that attract people to certain regions. Furthermore they take up the role of a symbol, standing for norms and rules shaped there or for a creative milieu due to their cultural diversity and thus establish trends through their cultural scene and capture public imagination as think tanks (Blotevorgel and Danielzyk, 2009; Feldman, 2005).

Due to the accumulation of important factors, which are carried in the streams inside these centers, there are certainly advantages in terms scale and network. Anyhow, in regards of innovation the most important fact is, that in these regions with a concentration of factors, knowledge spillovers do occur. Tacit knowledge is shared and spread throughout the region through fluctuation between companies and social life that brings together people of a common background. Even though relational proximity is sometimes considered more important than regional proximity, the combination of both aspects is the ultimate form of an innovation friendly environment (Kujath, 2009).

The flows of all the mentioned streams form what people call pulsating (Grabow and Becker, 2009). The region is vibrant and lively; it is a breathing and developing system (Grabow and Becker, 2009). This vivid explanation does also illuminate, why Cooke and Morgan refer to a metropolitan region as a nexus of processes. A region creates links and connections between processes that form a lively system. It thus creates a system that is capable of generating innovations (Cooke and Morgan, 1998). This is why global firms often establish their regional headquarters in metropolitan areas (Blotevorgel and Danielzyk, 2009).

All innovative regions have a starting point from which they grow. But why do they grow? Workers come to work, gain knowledge, maybe start up own business and attract more workers this way. The influx of people and new companies to a region creates an increase in the demand of tertiary sectors, such as shopping, housing and other services. Innovative clusters benefit from these reinforcing processes (Feldman, 2007). Another factor, why certain technologies tend to form in clusters are spin-offs, employees leaving their company to start up a business of their own. Most likely the business will be according to his or her skills, so in a sector like his old company (Klepper, 2007). In consequence, the likelihood of path dependency is quiet high. If a region has grown to a cluster of certain technologies or businesses, its fate is closely linked to the products typical for the region. So a product life cycle easily develops to be regional life cycle. This underlines the fact, that innovation is needed to sustain the wellbeing industry and the region (Thomas, 2005).

2.3   Services and intermediaries

Innovations are commercialised inventions and the knowledge communities are the starting point for innovations. Ideas are commercialised in already existing companies or start-up businesses. Start-ups mostly need certain forms of support like financing, legal service or management consulting. A regional nexus is needed to facilitate contacts and collaboration among potential business partners (Cooke and Morgan, 1998). Being embedded in a network or community means, they not only are visible to other stakeholders, like banks or venture capitalists that are interested in promising investments. Furthermore this embeddedness induces a form of mutual trust and a loss of anonymity. This altogether creates an environment of reliability, accountability and thus trustworthiness of the ideas and businesses and consequently makes it more interesting for investors and lenders (Storper and Venebales, 2005). Networking and interacting is a crucial force, which attracts individuals and firms into clusters, where skilled workers, other firms and formal and informal collaborations are present. A company's or individual's ability to interlink and to socialize and communicate within the local community is crucial for their success in taking part in innovation. They need to be embedded in a thick network, which is sharing knowledge (Breschi and Malerba, 2005). To systematically maintain the arrangement of contacts among potential partners, intermediaries are often necessary. These intermediaries like associations, institutions, formal networks or dedicated individuals are the nodes between the potential partners (Baxter and Tyler, 2007).

Up to now we mostly acted on the assumption, that individuals follow this process. In fact, the same applies to multinational enterprises that establish subsidiaries in certain areas. They gain access to local knowledge through there subsidiaries and get embedded in a local community (Blanc and Sierra, 2007).

So in wider scope, with linking their internal network with external and locally embedded knowledge networks, they create transnational spaces for learning. Japanese enterprises for example are running R&D laboratories in universities as intermediaries in the United Kingdom. The cooperation between the university and the company creates a win-win-situation. The company is embedded in the local scientific community, can employ students or scientists for scientific projects, while the universities have a partner to put their breakthrough technologies to market and make their share of earning through patents for instance (Lam, 2007).

2.4 Systematic innovation and institutional underpinnings

In order to create successful regional clusters of innovation it needs regional innovation systems. According to Cooke and Morgan they are characterised by regions, which possess the full panoply of innovation organizations set in an institutional milieu, where systematic linkage and interactive communication among the innovation actors is normal (Cooke and Morgan, 1998). Within the regional context it is up to the governments to provide underpinnings, which support successful entrepreneurial activity. These institutional underpinnings should follow the guiding question of how tacit knowledge is produced and which role the framework plays (Nonaka, 1991). This means, they have to satisfy the needs for supportive social capital (non pecuniary factors, that facilitate information flow), infrastructure, availability of venture capital or loans and entrepreneurial support services like lawyers, consulting or research institutions (Feldman, 2005). Regional institutions should foster cooperation among enterprises, add to business support, facilitate dialogue between local partners and reinforce university-industry cooperation (Cooke and Morgan, 1998). Bearing in mind the importance and the advantages of being embedded in a community, systematic innovation support is clearly handled better by sub central or regional institutions or governments rather than on national level, because they are acting within a network and are therefore aware of the needs of innovative start-ups.

Banks and venture capitalists seek profit and profit can only be made of a commercialisation of an invention. So mostly only the latter is attractive to them. As a consequence fundamental research needs to be funded by public sectors (Cooke, 2005). This concerns efforts in all regards. The primary task of innovation policy is facilitating knowledge transfer between subsystems and enhancing knowledge flows with the creation of scientific institutes, support of trade fairs or conferences or support of networks. Furthermore policy should also fund research projects and stimulate private investing organisations with a profit motive, so it has to be considered, that public institutions often have to give a public grant as a precondition for a loan from a bank (Cooke, 2005). In a green paper, the EU innovation and regional development named some key factors, which act as barriers for innovation such as cultural financial or regulatory factors and three key weaknesses. Firstly a proportionally less investment in R&D, secondly, little coordination between various stages of R&D and thirdly, limited capacity to convert innovation into commercially successful products (Cooke and Morgan, 1998). A systematic governmental approach is referred to as regional innovation systems, where close interfirm communication, socio-cultural and institutional environment may stimulate socially and territorially embedded collective learning and continuous innovation (Cooke, 2004).

3 European high-tech regions as empirical evidence

3.1 The Cambridge Phenomenon

During the 1970s the economic success of the region in and around Cambridge in the United Kingdom received broad attention all over Europe. A region, where there has merely been much industry before at all, grew to be a major business center for computing hardware and software, scientific instruments, electronics and telecommunications and increasingly biotechnology (Mawson, 1985).

What had happened? Many factors played key and reinforcing roles in causing and shaping the so-called Cambridge Phenomenon and essentially it was the result of several long-term and preconditioning factors reinforcing each other. Nevertheless there are certain factors, which make the Cambridge phenomenon so characteristic. Among the first to mention would be the venerable Cambridge University, which produced such great scientists like Sir Isaac Newton, Ernest Rutherford or Niels Bohr, since long enjoyed a high reputation in enterprise, research and technology. It has exercised a profoundly important influence, directly but especially indirectly (Segal, 1985). A son of Charles Darwin, who was a member of university's constituent Trinity College, founded Cambridge Instruments in 1881 (Mawson, 1985), one of the first technology companies in the region. A few key individuals in the University, influenced by what they saw happening around Stanford and MIT, perceived that the vitality, relevance and funding of the University's research would be dependent upon a good number and diversity of science-based companies and of non-academic research establishments being close to the University. They also saw, that research students would increasingly have to find careers in industry not academe, which would be facilitated by the proximity of science-based industry. This would make it easy for faculty members to enter into commercial activity and to transform their inventions into innovation with commercial success (Mawson, 1985). As an initiative of the university's constituent Trinity College the Cambridge Science Park, a cluster of small companies, sometime as little as 3 employees, often being start-ups of scientists or graduates from the Cambridge University, was founded in 1970. Consequently one can say, that the decision of the Trinity College was the very origin of virtually all the companies now present in the area (Mawson, 1985).

To facilitate contacts between individuals or several networks like the Cambridge Network were set up. Today there are numerous interlocking networks of talented, influential and accessible individuals, which make for informal, congenial and efficient business dealings (Mawson, 1985). Even though the Cambridge Science Park has come to play a critical role in sustaining and enhancing the phenomenon, the trigger to the phenomenon was rather a spirit than infrastructural environment (Mawson, 1985). The development spread throughout the region outside of the Cambridge Science Park. More science parks were set up in the region, but the high-tech sector not only grew in the science parks. All over the region small scale high-tech companies were established.

One key service factor for commercial success was finance and business services, the underpinnings of successful business. The economic success of the Cambridge region can be substantially traced back to a strategic decision of Barclays Bank in the 1970s to invest money and time in the business plan of first-time technological entrepreneurs. The first signs of success of this strategy encouraged other forms or financial institutions operating nationally and internationally, as well as other local investors, to finance the local high technology companies both at start-up and later rounds of financing (Mawson, 1985). The Barclays Bank decision strengthened confidence of accountants, solicitors and others in the local business community in getting involved supporting such enterprises even in an environment, where economic success was of high uncertainty. As a consequence today many companies covering technology, but also peripherical sectors like finance and services can be found throughout the region. Those firms were playing an important role in stimulating further development in the local industrial and commercial sectors. Second, an industrial milieu was created. The fact that there has never been heavy industry, or industries in which large plants and large unionized labour forces have been prominent has helped create a labour market and a general attitude in which flexibility and individualism have never been suppressed.

In the beginning Cambridge did not offer infrastructural special features. There were conventional housing and business buildings. This shows, that planned provision of property is not itself a sufficient factor and sometimes not even a necessary factor in stimulating development of high technology industry. Research scientists live in the area, and there is a continuing vitality about the local cultural and social scene. The University has created a unique environment for social and inter-disciplinary contact within the entire academic and research community and also extending to the local high technology and business communities. This way the University created a 'culture of excellence and openness' (Mawson, 1985). In essence the University has a supportive posture towards faculty members' involvements of all kinds with industry (Mawson, 1985). With all the new start-ups and other companies establishing subsidiaries in the Cambridge region, the number of skilled workers and scientists grew steadily. Some were recruited directly from university and thus stayed in the region and some came to the region from elsewhere, so that the scientific community developed and grew significantly. In this context Mawson is talking about the phenomenon of people being 'sucked into' Cambridge as for skilled labour and scientists are attracted (Mawson, 1985). In the following years Cambridge developed to be a showcase for innovative regions. Today the region of Cambridge is home to a large number of high technology businesses. Recent estimates lead to the figure of 1000 firms in 2007 compared to some 400 in 1984 (Mawson, 1985; Library House, 2007). Most of these businesses are small and often have less than 30 employees (Segal, 1985). So one can say, that the Cambridge phenomenon is being driven by small local enterprises and that other local resources like the University, banks, business community and so on – are closely related to the success (Segal, 1985). Typically the high-tech companies, that are found in and around Cambridge are set up by individuals or small groups, which were 'spinning out' of existing firms and other local organizations including the University. Principally the mainly young start-ups are engaged in research, design and development of products of low quantity, but of high value technological production (Mawson, 1985).

Figure 3 shows that venture capital investments in the first half of 2007 were highest in the London and Cambridge regions. This verifies the outstanding economical role Cambridge plays together with the London region in the British economy.

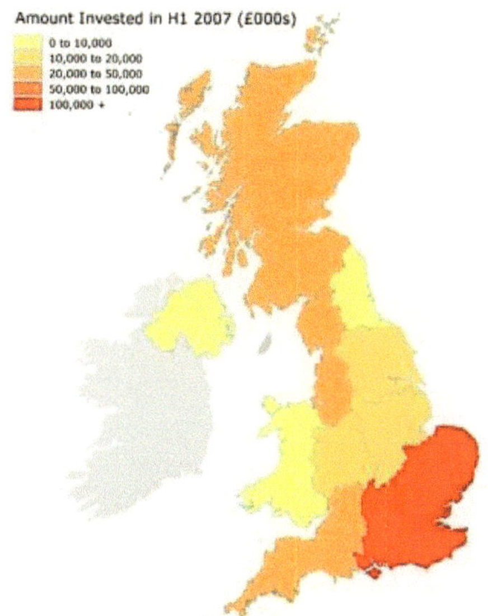

Figure 2: Venture capital investments in H1 2007 in government regions
source: Cambridge Cluster Report 2007

The Cambridge region is often referred to as Silicon Fen to draw parallels to the Silicon Valley in the United States and to emphasize its scientific importance in the Fenland. It does now serve as a visible and prestigious symbol to the outside world. Nevertheless the Cambridge phenomenon is specific to Cambridge to a certain extend. Many attempts to copy it elsewhere were of little or no success. Thus, one has to emphasise, that due to its very special environment the Cambridge case is not a model to be slavishly followed in other locations, but rather a trigger to think about the local settings, which favour innovative activity in certain regions (Segal, 1985). The principle of university-linked high technology development is now under way in many locations. Universities in many places enforce links to industry, like common projects or the integrated chairs in Germany (Mawson, 1985).

3.2   Munich – Leading high tech region in Germany

Since long Munich is in an outstanding position as a high technology region in Germany. The region of Munich is home to large enterprises like Siemens, BMW or the German headquarters of EADS and to a vast number of small and medium sized enterprises (SMEs). Industries usually are dispersed throughout Germany, especially in the region of Munich, they are concentrated to a noticeable extend. This concentration favours interactive technological business cooperation between SMEs and the large enterprises. Within the region the city of Munich is clearly the center of the high technology community (see Figure 3). Approximately 70% of all high technology employees of the region were concentrated in the city of Munich by the end of the last century and by 1996 76,2% of all Munich manufacturing employees were employed in high tech sectors. So all in all, Munich maintains a strong specialization in the high technology sector (Sternberg and Tamàsy, 1998).

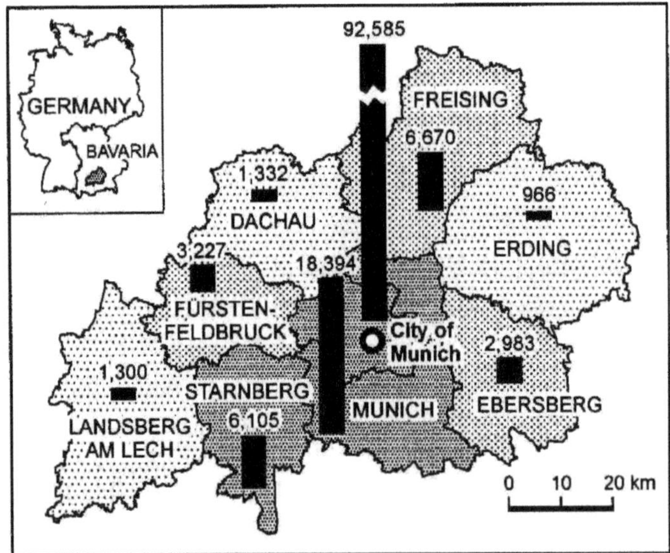

Figure 3: Distribution of employment in technology industries in Munich
Source: Sternberg and Tamàsy, 1998 Bavarian Statistical Office, 1997.

Historically two major events had significant influence on Munich successful development. Firstly, due to some military policy after the Second World War the Siemens headquarters were relocated from Berlin to Munich in 1948. Focusing on the role of Siemens within the region of Munich, the importance of this event becomes clearer. Siemens today is a global player in multitechnology with its largest R&D center being located in Munich. Siemens acts as supplier, customer and co-operator in microelectronic and other industrial sectors for other companies, which are mainly SMEs. It also acts as an enforcing node between several SMEs. When two companies for example are involved in a Siemens project, the ties between them might strengthen in terms of cooperation in order to increase their performance and competitiveness and thus be an attractive business partner for Siemens. Furthermore the large firms have considerably high financial resources and need knowledge of SMEs, which results in an interlinked sectoral cluster around the large enterprise (Sternberg and Tamàsy, 1998). Secondly, major political decisions in the 1950s and 1960s regarding the location of R&D institutions favoured Munich, influenced long term Bavarian Prime Minister and Federal Defence Minister Franz-Josef Strauss. Today the region Munich is best served by state funded research projects in Germany. It is home to 13 of 79 Max-Planck Institutes and it gets a major share of R&D funding by the Federal Ministry for Education, Science, Research and Technology (Sternberg and Tamàsy, 1998). This together with the educational institutions of the region provides adequate underpinnings for innovative activity. A large number of students and academic employees at the widely renowned universities in and around Munich create a vast amount of qualified workforce. With its positive location image, Munich additionally attracts people from elsewhere, which results in a highly specialised job market and a creative and innovative milieu. Additionally business spin offs of local expertise from large firms like Siemens or BMW do occur and result in a wide range of new SMEs, that are fuelled by informal contacts to former colleagues, facilitating sharing of knowledge and information (Sternberg and Tamàsy, 1998). This accumulation of knowledge, by generating new knowledge, keeping knowledge and attracting new knowledge to the region makes Munich a flourishing place of innovation.

3.3 Baden Württemberg – A successful region facing challenges

The region of Baden Württemberg is home of many of the worlds most renown engineering companies. This is especially the case for the automotive sector, with companies like Daimler, Bosch and Porsche, but as well for other engineering sectors such as aviation and machinery. The region is considered far more innovative than the European average (Cooke, 1997).

Historically mainly poor farmers inhabited the region, but in the late 19th century the region was pretty much the birthplace of the automobile, with Gottlieb Daimler, Rudolf Benz or Robert Bosch being among the inventors of breakthrough technology. After the Second World War it rose from a chaos and was the destination of many immigrants from eastern Germany as well as from other place in Europe like Turkey, Italy or Portugal. Many of those companies were and still are referred to as the so called "Mittelstand", small and medium sized companies, even though a clear line cannot be drawn as for Bosch refers to itself as Mittelstand as well. This "Mittelstand" together with the major companies played a key role in the German "Wirtschaftswunder" (Cooke and Morgan, 1998). Significant for the whole region is a very strong infrastructure of business support. The chamber of commerce and industry, several other government agencies and cooperative initiatives do have significant influence and a very strong standing. They promote networks among SMEs. The same applies to the unions and business associations. They are alliances under which roof many companies are active and they are the linking element between firms that seek cooperation or other contacts to potential business partners in terms of collaboration. These entire institutions act as intermediaries. The University of Karlsruhe with a famous engineering faculty is located in the region. But there is significant university-industry cooperation with the other German top universities in engineering such as the RWTH in Aachen or the TU Darmstadt, which are not located in Baden-Württemberg. This somehow questions the necessity of a university being present in the area, but rather having a strong relational link. Baden-Württemberg is path dependant to a large extend, as for automotive and machinery providing the biggest share of its economy. There is not so much of what one would call breakthrough innovative technology, but rather innovation with a direct link to automotive and machinery. The question is, whether the region needs to break out of this path dependency or, if the metamorphosis of the automotive sector in terms of alternative engines like fuel cell or hybrid provides sufficient ground for further innovation and thus the development of the region (Cooke, 1997).

3.4 Tampere – From resource-based to knowledge-based economy

The Finnish economy has been dominated by forestry and forest-related industry since the early stages of industrialisation and many factors in Finnish economy adapted to their needs. The economic crisis in the 1990s lead to a massive drop in exports and production and consequently to a rapid increase of unemployment with rates up to 20%. Within the crisis it was seen, that only a breakout of this path dependency would help to overcome the crisis and avoid a lock-in effect. Finland needed to transform from a resource-based to a knowledge-based economy (Schienstock et al, 2004).

The region of Tampere in southwest Finland accounts for some 11% of the total Finnish output and is the second most important traffic node after Helsinki. It was the founding place of Nokia and today is a center for information and communication technology (ICT) in Finland with annual growth rates in the ICT sector of up to 30% by 2004 (Schienstock et al, 2004). The case of Nokia might be exemplarily for Finland's economic transformation as for it was founded as a producer of paper and later rubber products, just to develop to be one of the world's leading ICT companies. Today the company is the biggest mobile phone producer on earth and acts as the engine for the development of the Tampere region, which now has a diverse structure with telecommunications, telecommunication services, production, R&D, software, information system design and other businesses related to ICT. In the recent years Nokia mostly accounted for nearly half the growth of the Tampere Region ICT sector, resulting in the Tampere region being second after Helsinki in patent applications and R&D expenditure (Schienstock et al., 2004). The ICT sector is the cornerstone in industrial renewal in Tampere, together with other sectors like new media or health care, an innovation system, which includes life science, pharmaceuticals, biotechnology and patients and consequently covering a major share of the GDP (Cooke, 2004). As one of the industrial centers in Finland the Tampere region was hit severely by the economic crisis of the 1990s and as it turned out, electronics was leading the way to industrial renewal away from resource-based industry. Traditionally Finland was a very centralistic administration, but the local authorities in Tampere enjoyed remarkable autonomy in its decisions. This freedom of scope was used to initiate several programmes promoting innovation and to open regional government offices, that bring together functions of several ministries providing advisory services, specific sector information, consultancy services, training and finance services (Regional Employment and Economic Development Center of Tampere EEDC). Intermediaries like science parks, research facilities, science institutes and the Tampere Center of Expertise programme were set up to strengthen the link between science and local business actors and to facilitate technology transfer (Schienstock et al., 2004).

Another favouring factor for the region's successful development were the education institutions, which provide valuable research capabilities and highly qualified workforce. There are several polytechnics, research laboratories and two universities, one of which is the Tampere University of Technology (TUT) with a center for semiconductors enjoying research cooperation with Nokia. The intermediary system actually works so well, some 80% of all knowledge intensive firms, that employ ten or more people are involved in cooperation with educational organisations in region (Schienstock et al., 2004). The often noticed phenomenon of spin-offs from big companies is most likely also the case for former Nokia employees founding new firms.

Besides private venture capital companies, that expanded their activity in the 1990s a wide range public financial support programmes was put in place. The National Technology Agency (Tekes) for example provided R&D grants, the state-owned credit institution Finnerva offered short-term and long-term loans for SMEs and even the council of the region and the city of Tampere act a financer and shareholder (Schienstock et al., 2004).

Very often Nokia is involved R&D partnerships with big and small firms or subcontracts R&D activities. So Nokia can be seen as the magnet, that attracts companies and people with capabilities in ICT, that want to participate. With the rapid development of the ICT market towards the end of the 1990s, more companies were set up or moved to the region in order to be embedded in the ICT community and have access to the regions knowledge, potential partners, customers, suppliers and supporting services. Consequently, the region built up a diverse social and cultural landscape, a knowledge pool and a specialised job market. One might doubtless say, that Tampere's development would have been significantly different, if it weren't for the fact that Nokia as a global market leader in ICT was already there. Nevertheless, the key to the remarkable economic success of the region was the provision of an appropriate framework by the local authorities, which attracted companies and people, supported entrepreneurs and supported product innovation.

3.5 Grenoble – France's leading edge technology center

Grenoble in the French Rhône-Alpes region is a major scientific center for physics, computer science and applied mathematics and is considered to have one of the most innovative milieus in Europe (Rousier, 1993; Sternberg, 1996). The development of the region can be traced back to the Nobel Prize laureate in physics, Professor Louis Neel. His influence and efforts lead to the national Atomic Energy Commission (CEA) and the National Energy Research Center (CEN) being decentralized and established in Grenoble in 1956 (Sternberg, 1996). This event can be considered as the trigger for the Grenoble region becoming a major technological center, employing some 21.000 people in research, more than half of them in public research. The region is today known as the "filière électronique" and home to a large number of companies dedicated to electronics, like STMicroelectronics or Hewlett Packard France. Some 60.000 students are educated in the city's universities with the Joseph Fourier University (UJF) being a leading French natural scientific university. As part of the university, the Grenoble Institute of Technology (INPG) trains more than 5.000 engineers every year in key technology disciplines. Many fundamental and applied scientific research centers laboratories like the European Synchrotron Radiation Facility (ESRF), the Institut Laue-Langevin (ILL), the European Molecular Biology Laboratory (EMBL) and one of the Commissariat à l'Énergie Atomique (Nuclear Energy Commission CEA) main research facilities are located in the region Sternberg, 1996).

Since long Grenoble is characterised by a strong link between the local research and education institutions and the SMEs. As one of the first technology parks in France the Zone pour l'Innovation et les Réalisations Scientifiques et Techniques (ZIRST) was founded in 1972 by the university and state research institutions. The ZIRST was an early systematic approach to facilitate contacts between sciences and business, technology transfer is seen as an incubator for many businesses in the region (Chanaron and Ruffieux, 1990; Sternberg 1996). In 2005 the center was renamed Inovallée to create a catchy synonym, that implies innovation and technology like the term "Silicon Valley".

The decentralization of state R & D institutions, which are usually located in Paris, such as the CEN, and their establishment in the Grenoble region gave important impulses for the regions development into a technopolis of this order of magnitude (Dunford, 1989; Sternberg 1996). The early 1980s were of particular importance in this context. At that time, government technology policy heavily invested in electro technology (in particular semiconductor technology)

The divers scientific community lead to a cultural diversity, which is enriching the vibrant scene of the city. This in conjunction with the picturesque setting is another pull factor, which will continue to attract skilled labour into the region and thus enrich the knowledge pool.

4   Conclusion

Even though the findings in chapter 2 are based on empirical studies and are thus somehow plausible, the examples for innovation regions given do verify a key point yet another time. Proximity matters. The regions that have been examined more closely in this paper do share an important aspect: Their success is based on the geographical and relational proximity of the actors. The socio-economic dynamics tend orientate themselves to a nod. The factors contributing to a successful innovation do need a regional nexus, which brings them together. Participating means being, where things happen and taking part in the events. As for innovation mostly promises profit, everyone, who wants to participate in this profit, has to 'be there'. He is seeking knowledge, opportunities and support, most of what he only finds in close proximity to the innovation center.

Often it only needs the local setting of a key institution like the Cambridge University, of the Siemens headquarter or defence R&D institutions in Munich, the inventor of the automobile in Baden-Württemberg, Nokia in Tampere or central scientific centers and commissions in Grenoble to attract certain technologies to a region. Policy in the best case can trigger a regional innovation cluster, but most likely has the task to provide the proper environment for already existing sparks of innovation to flourish. Nevertheless, as the example of Baden-Württemberg shows, every innovative region is subject to path dependency to a large extend and consequently has to reinvent itself from time to time to avoid lock-in effects.

REFERENCES

Amin, A. and Cohendet, P. (1999): Learning and adoption of decentralised business networks, in: Environment and Planning D – Society and Space 17, 87-104.

Audretsch, David B and Feldman, Maryann P. (1999): Innovation in Cities: Science-Based Diversity, Specialization and Localized Monopoly, in: European Economic Review, Vol. 43, 409-429.

Baxter, Christie and Tyler, Peter (2007): Facilitating enterprising places: The Role of Intermediaries in the United States And the United Kingdom, in: The Economic Geography of Innovation, edited by Karen R. Polenske, Cambridge University Press Cambridge/New York, 261-288.

Blanc, H. and Sierra, C. (1999): The internationalization of R&D by multinationals: A trade-off between external and internal proximity, in: Cambridge Journal of Economics, 23, 187-206.

Blotevogel, Hans Heinrich and Danielzyk,, Rainer (2009): Leistungen und Funktionen von Metropolregionen, in: Metropolregionen – Innovation, Wettbewerb, Handlungsfähigkeit, Forschungs- und Satzungsberichte der ARL Metropolregionen und Raumentwicklung Teil 3, edited by Jörg Knieling, Akademie Raumordung und Landesplanung, Hannover, 22-29.

Breschi, Stefano and Malerba, Franco (2005): Clusters, Networks and Innovation: Results and New Directions, in: Clusters, Networks, and Innovation, edited by Stefano Breschi and Franco Malerba, Oxford University Press, New York, 1-26.

Chanaron, J. J. and Ruffieux, B. (1990): The Efficiency of Technopoles, Lessons from the Experience of Meylan's ZIRST, in: Industry and Higher Education, June, 131-135.

Cooke, Philip (1997): Regions in a global market: the experiences of Wales and Baden-Württemberg, in: Review of International Political Economy, 4:2, 349-381.

Cooke, Philip and Morgan, Kevin (1998): The associational economy – firms, regions, and innovation, Oxford University Press, Oxford/New York.

Cooke, Philip (2004): Regional innovation system – an evolutionary approach, in: Regional Innovation Systems, 2nd edition – The role of governance in a globalized world, London/New York, 1-17.

Cooke, Philip (2005): Regional knowledge capabilities and open innovation - regional innovation systems and clusters in the asymmetric knowledge economy, in: Clusters, Networks, and Innovation, edited by Stefan Breschi and Franco Malerba, Oxford University Press, New York, 80-109.

Dunford, M. (1991): Industrial Trajectories and Social Relations in Areas of New Industrial Growth, in: Industrial Change and Regional Development, edited by G. Benko and M. Dunford, Belhaven, London, 51-82.

Feldman, Maryann P. (2005): The entrepreneurial event revisited - Firm formation in a regional context, in: Clusters, Networks, and Innovation, edited by Stefan Breschi and Franco Malerba, Oxford University Press, New York, 136-168.

Gertler, Meric S. (2007): Tacit Knowledge in the production systems: how important is geography?, in: The Economic Geography of Innovation, edited by Karen R. Polenske, Cambridge University Press Cambridge/New York, 87-111.

Grabow, Bussow and Becker, Anna (2009): Metropolregionen - Quellen und Mündungen von Wanderungsströmen, in: Metropolregionen - Innovation, Wettbewerb, Handlungsfähigkeit, Forschungs- und Satzungsberichte der ARL Metropolregionen und Raumentwicklung Teil 3, edited by Jörg Knieling, Akademie Raumordung und Landesplanung, Hannover, 270-299.

Jaffe, A. B., Trajtenberg, M. and Henderson, R. (1993): Geographic localization of knowledge spillovers as evidenced by patent citations, in: Quarterly Journal of Economics, 108, 577-598.

Klepper, Steven (2005): Employee start ups in high tech industries, in: Clusters, Networks, and Innovation, edited by Stefan Breschi and Franco Malerba, Oxford University Press, New York, 199-231.

Kujath, Hans Joachim (2009): Leistungsfähigkeit von Metropoloregionen in der Wissensökonomie – Die institutionentheoretische Sicht, in: Metropolregionen – Innovation, Wettbewerb, Handlungsfähigkeit, Forschungs- und Satzungsberichte der ARL Metropolregionen und Raumentwicklung Teil 3, edited by Jörg Knieling, Akademie Raumordung und Landesplanung, Hannover, 200-222.

Lam, Alice (2007): Multinationals and transnational social space for learning: knowledge creation and transfer through global R&D networks, in: The Economic Geography of Innovation, edited by Karen R. Polenske, Cambridge University Press Cambridge/New York, 157-189.

Library House (2007): Looking Inwards, Reaching Outwards – The Cambridge Cluster Report 2007, Cambridge.

Lundvall, Bengt-Ake, Johnson, Björn, Andersen, Esben S. and Dalum, Bent (2007): National systems of production innovation and competence building, in: The Economic Geography of Innovation, edited by Karen R. Polenske, Cambridge University Press Cambridge/New York, 213-240.

Malmberg, A., Sölvell, Ö. and Zander, I. (1996): Spatial clustering, local accumulation of knowledge and firm competitiveness, Geografiska Annaler 78 B, 85-97.

Maskell, Peter (2005): Towards a knowledge base theory of the geographical cluster, in: Clusters, Networks, and Innovation, edited by Stefan Breschi and Franco Malerba, Oxford University Press, New York, 411-432.

Mawson, J. (1985): Policy Review Section, Regional Studies, 19:6, 563-578.

Nonaka, Hirotaka (1991): The knowledge creating company, 1991, in: Harvard Business Review 69; 96-104.

Rousier, N. (1993): Grenoble-A Technological District? In: Technopole Planning in Britain, Ireland and France: The National Technology Policies-The Planned Regional Acceleration of Innovation, edited by Simmies, J., Cohen, J. and Hart, D. Planning and Development Research Center, Universitiy College London, London, 115-120.

Schienstock, Gerd, Kautonen, Mika and Koski, Pasi (2004): Escaping path dependency – The case of Tampere, Finland, in: Regional Innovation Systems, 2nd edition – The role of governance in a globalized world, London/New York, 127-153.

Segal, Quince and Partners (1985): The Cambridge Phenomenon: The Growth of High Technology Industry in a University Town, Cambridge.

Sorensen, Olav (2005): Social networks and the persistence of clusters: Evidence from the computer workstation industry, in: Clusters, Networks, and Innovation, edited by Stefan Breschi and Franco Malerba, Oxford University Press, New York, 297-316.

Sternberg, Rolf (1996): Reasons for the Genesis of High-Tech Regions – Theoretical explanation and empirical evidence, in: Geoforum, Vol. 27. No. 2, 205-223.

Sternberg, Rolf and Tamàsy, Christine (1998): Munich as Germany's No 1 High Technology Region: Empirical Evidence, Theoretical Explanation and the Role of Small Firm/Large Firm Relationships: Regional Studies, 33:4, 367-377.

Sternberg, Rolf and Krymalowski, Mark (2002): Internet Domains and the Innovativeness of Cities/Regions – Evidence from Germany and Munich, European Planning Studies 10:2, 251-273

Storper, Michael and Venables, Anthony J. (2005): Face-to-Face contact and the urban economy, in: Clusters, Networks, and Innovation, edited by Stefan Breschi and Franco Malerba, Oxford University Press, New York, 319-342.

Takeuchu, Ikujiru and Nonaka, Hirotaka (1995): The Knowledge-Creating Company: How Japanese Companies Create the Dynamics of Innovation, Oxford University Press.

Thomas, Morgan D. (2005): Growth Pole Theory: technological change, and regional economic growth, in: Papers of the Regional Science Association, 43, 3-25.

Other resources:

Siemens AG:
http://w1.siemens.com/innovation/de/ueber_funde/corp_technology/standorte.htm, December 2009

AEPI - Grenoble-Isère economic development agency:
http://english.grenoble-isere.com/195-research-grenoble-isere-rhone-alps.htm)